Process Development Division 2022

Held at the 2022 AIChE Spring Meeting and 18th Global Congress on Process Safety

San Antonio, Texas, USA
10 – 14 April 2022

ISBN: 978-1-7138-6829-3

Printed from e-media with permission by:

Curran Associates, Inc.
57 Morehouse Lane
Red Hook, NY 12571

AIChE
120 Wall Street, FL 23
New York, NY 10005-4020

Phone: (800) 242-4363
Fax: (203) 775-5177

www.aiche.org

Additional copies of this publication are available from:

Curran Associates, Inc.
57 Morehouse Lane
Red Hook, NY 12571 USA
Phone: 845-758-0400
Fax: 845-758-2633
Email: curran@proceedings.com
Web: www.proceedings.com

TABLE OF CONTENTS

640286 Scale-up of Valor® Coated Glass Vials for the Pharmaceutical Industry

Abstract:

Over the last 10 years Corning has applied expertise in glass science and other core technologies to develop Valor® Glass— a purpose-built pharmaceutical glass specifically designed to address the challenges of today's manufacturing operations. With the onset of the COVID-19 pandemic, the demand for these Valor® Glass vials increased dramatically due to their higher filling line throughput and improved damage resistance during cold storage compared to standard borosilicate vials. The vials are chemically strengthened through ion exchange and a low coefficient of friction coating is applied to the exterior surface to improve machinability. With increased vial demand came the need to rapidly increase capacity, so Corning built a new factory in Durham, NC with production beginning in Fall of 2021. This presentation will discuss the best-practices and key learnings that enabled the timely and cost-effective scale-up of the manufacturing processes used to produce Valor® glass vials.

640272 Utilizing Azeotropes & Solvent Recycle for Cost Effective Implementation on Batch 1

Abstract:

This talk describes the development and scale-up of a solvent recovery strategy that utilized fundamental experimental data to support successful commercial implementation. The commercialization of an agrochemical intermediate required solvent recovery within a triethylamine-acetonitrile-water solvent system. The design of the recovery process was inhibited by the lack of reliable experimental data for this mixture. A spinning band column was used to determine the equilibrium azeotropic compositions. A successful process was developed based on these data to maximize the recovery of all three species. Additionally, the implementation of the solvent recovery improved the downstream crystallization and reduced the amount of waste generated. These detailed studies proved invaluable during start-up and were required on the very first batch due to an unforeseen global shortage of acetonitrile.

639934 Process Development Tools to Maximize Research Efficiency

Abstract:

Research is often challenged to move faster and spend less time in the lab to increase the speed to the market. However, there are times where a little more time in the lab generates better outcomes for customer satisfaction and product quality. This work will highlight some key tools such as online analytical tools, automated parallel reactors, and Design of Experiments that were used for process development of an organosilicone compound. By utilizing the correct tools in specific stages of commercialization, the balance between research quality and speed to the market was optimized.

640329 Hydrogen based processing of mineral iron carbonate, process and economics

Abstract:

Hydrogen based reduction of mineral iron carbonate ore is expected to contribute to carbon dioxide emission reduction. At present the annual global iron and steel production capacity is about two billion metric tons, with specific carbon dioxide emissions of about 1.9 to 2.2 metric tons per ton of pig iron. Hydrogenation of carbon dioxide (CO_2) may contribute to the reduction of CO_2 emissions. High carbon dioxide concentration of process gas from direct iron ore reduction with hydrogen recommends conversion of carbon dioxide to either methane or methanol.

Catalytic conversion of carbon dioxide to methane with a MgO/Ni catalyst was successfully investigated in lab scale. The product gas may directly be fed into a gas grid. The spent catalyst may easily be recycled in the steel industry.

The bulk chemical methanol is a suitable alternative product. It is used for paints, plastics, and adhesives, as feedstock for the production of chemicals, and as fuel additive or alternative fuel. It provides promising properties for hydrogen storage, as it can raise the energy density of hydrogen-based energy carriers by one order of magnitude. Low temperature level in reforming also suggests methanol for fuel cell applications. State-of-the-art methanol production is based on syngas, a mixture of carbon monoxide, CO_2, and hydrogen, and preferably $Cu/ZnO/Al_2O_3$ catalysts. State of the art reactor- and process design suggest looping systems with sophisticated internal heat energy management. Huge efforts have meanwhile been made to directly process carbon dioxide instead of syngas. To our knowledge neither of carbon dioxide based processes has established on the market, mainly because of catalyst life time limitations. To sum up the present status, the establishment of direct methanol synthesis from carbon dioxide suffers from limitations in feed gas preparation and/or appropriate catalysts. In process gas from direct reduction of mineral iron carbonate carbon dioxide at high concentration is available in huge quantities. When processing carbonate based iron ores (siderite) in hydrogen atmosphere our group registered that at ambient pressure within a temperature window of 600 to 800 °C it is possible to alter the carbon monoxide/carbon dioxide ratio of the exiting gas. Methanol synthesis with off-gas from mineral iron ore reduction with hydogen was successfully tested and validated in a lab size CSTR loop reactor system with external cooling loop for condensing the gaseous products. The methanol synthesis process was validated with a Cu/MgO catalyst prepared by wet impregnation and stoichiometric CO_2/H_2 feed at 300 °C and 5 MPa.

The technology of pig iron production from mineral iron carbonate with hydrogen and carbon dioxide utilization has been successfully validated, but it has to be considered at what price we can establish pig iron production without carbon dioxide emissions.

According to the state of the art blast furnace process the specific energy consumption is 17.8 GJ/t pig iron, and the specific carbon dioxide emissions is about 1.9 t/t of pig iron.

Based on hydrogen production by electrolysis (10,000 A/m^2 at 1.84 V for zero gap electrolysis setup) the iron ore reduction step ($FeCO_3 + H_2 = Fe + CO_2 + H_2O$) will consume as much as 7.26 GJ/t of pig iron (plus 0.39 GJ/t pig iron for the smelter). Conversion of carbon dioxide to methane ($CO_2 + 4\ H_2 = CH_4 + 2H_2O$) will consume 29 GJ/t pig iron. Alternatively, conversion of carbon dioxide to methanol

($CO_2 + 3\ H_2 = CH_3OH + H_2O$) would consume 21.8 GJ/t of pig iron. Even when considering the LCV of methane and methanol the specific energy consumption for hydrogen based iron ore processing would consume at least 21.3 GJ/t of pig iron in case of methane synthesis and 22.9 GJ/t of pig iron in case of methanol synthesis. In conclusion, the technology of direct reduction of mineral iron carbonate with hydrogen is feasible. The gas quality of process gas from direct reduction of mineral iron carbonate with hydrogen permits utilization of carbon dioxide, but at the expense of the specific energy consumption.

637659 Early-Stage Strategy and Planning for Achieving Cost-Effective Commercialization of New Processes

Abstract:

The journey from process discovery to commercial deployment of any new process involves the intermediates steps of process development, piloting and technology demonstration. Commercialization cost depends on the size of the intermediate pilot and demonstration units, the type of experiments and duration of these experiments in these plants. Early envisioning of a final commercial process scheme and plant capacity helps to efficiently identify the right size and kind of these intermediate units. Also, early-stage identification of technology risks enables the planning of experiments required for adequate technology derisking. In addition, understanding and analyzing organizational skills and availability of in-house resources can help identify the need for collaboration and external third-party resources that can be leveraged during the journey. This presentation will address the influence of these factors in achieving cost effective new process commercialization.

639323 Waste Plastics to Cleaner Fuels: Technical and Economic Assessment

Abstract:

An increase in energy demand in the recent decades has caused energy shortages. An increase in the fuel prices has forced the humanity to use the non-conventional energy sources for heat, power and clean fuel production. On the other hand, the waste generation has also increased with an increase in the population. Gasification and partial oxidation are among the few techniques that has a potential to convert the solid and liquid waste into syngas that can be transformed into hydrogen and other fisher tropsch fuels. Polyethylene (PE) and polypropylene (PP) are considered in this study because of their massive worldwide availability in the category of waste plastics. Simulation models are developed in Aspen Plus (R) to convert the PE and PP into syngas using the gasification process followed by the integration of gasification unit with the methane reforming unit to enhance the overall syngas production capacity. The technical and economic assessment is performed to convert the syngas into H2 and Methanol. This process integration approach not only enhances the overall H2 and methanol production rates but also reduce the reliance on the single fuel. It has been seen from results that the new integrated design containing both gasifier and reformer enhances the hydrogen mass production rate per mass feed rate by 7.5%. Furthermore, the process performance analysis showed that the efficiency of the new process is increased up to 2.5%, where the hydrogen production cost showed the reduction of 30% compared to the standalone gasification case. The results also showed that the new process has a potential of reducing the carbon dioxide emissions by 5.0%.

639524 A New Conceptual Design on Natural Gas Comprehensive Utilization for Simultaneous Power, Ammonia, Fertilizer, and LNG Productions with Inherent Carbon Capture

Abstract:

Natural gas is the one of the most promising large-scale alternative energy/material resources in the world, which needs to be comprehensively utilized to satisfy the increasing demand on global energy/fertilizer consumptions with carbon sequestrations. This paper, based on our previous work [1, 2], has developed a novel conceptual design of an industrial complex to simultaneously generate power, ammonia, nitrogen fertilizer, and LNG; meanwhile, with its inherent carbon capture process to produce the pipeline-grade CO_2 product for many potential uses. It employs natural gas and air as the major feeds, and retrofit five industrial process sub systems including (i) an air separation unit (ASU); (ii) an Allam-cycle power plant (ACPP); (iii) an ammonia plant (AP); (iv) a urea fertilizer plant (UFP), and (v) an LNG plant (LNGP). Through optimally integrating energy/material streams among five subsystems, this new complex can simultaneously produce electricity, ammonia, urea, LNG, pure N2, and pipeline-ready CO2 products efficiently and effectively with the eco-friendly manufacturing ways that would prevent CO2 and NOx emissions. The newly developed complex has been modeled and examined through a commercial simulator, and its exciting economic and environmental performances have also been evaluated and demonstrated. The study has a great industrial potential with salient economic, environmental, and societal benefits, which could be further optimized in the future with possible interested industrial partners.

Keywords: Allam-cycle power plant; urea fertilizer plant; LNG plant; inherent carbon capture; process modeling

Reference:

1. Fernandes, D.; Wang, S.; Xu, Q.; Buss, R.; Chen, D. Process and carbon footprint analyses of the Allam cycle power plant integrated with an air separation unit. *Clean Technologies*, 1, 325-340, **2019**a.
2. Fernandes, D.; Wang, S.; Xu, Q.; Chen, D. Dynamic simulations of the Allam cycle power plant integrated with an air separation unit. *International Journal of Chemical Engineering*, 1-10, **2019**b.

639603 Design and Techno-Economic Analysis for the Lignin Oil Production Process from Lignocellulosic Biomass

Abstract:

Lignocellulosic biomass is an important feedstock for the biofuel and pulp and paper industry; it is also considered a promising source for producing many other high-value chemical products. Kai et al. (2021) has introduced a novel chemical process using lignocellulosic biomass to produce profitable products such as xylooligosaccharides (prebiotics), xylose, microfiber. In that process, lignin from the biomass (around 20% of the mass) is sent to the CHP plant. However, there is a great potential to improve the plant profitability further. A valuable product, i.e., lignin oil, can be obtained from lignin and is expected to enhance the economic benefit significantly. We have developed such technology to extract lignin oil through isopropyl alcohol without any catalyst. This lignin oil product can be used as a cosmetic additive since it provides good UV protection for human skin. With an increasing cosmetic market, lignin oil will have good business prospects.

However, to commercialize such technology, a process must be developed economimically. This study will develop the plant-scale lignin oil production process, find the best design plan, and analyze the economic performance: (i) A series of lab experiments will test the thermodynamic behavior of lignin oil. (ii) Based on the properties, several plant-scale process designs will be developed, and the process feasibility will be examined with rigorous mass and energy balance in Aspen Plus. (iii) Preliminary economic analyses will be performed for each process design in order to find the best option. (iv) A techno-economic model with discounted cash flow rate of return (DCFROR) calculations is developed for the best design; sensitives to the economic performance will also be discussed.

This work will provide valuable information for the future commercialization study on this technology.

Reference: 1.Lan, K., Xu, Y., Kim, H., Ham, C., Kelley, S.S. and Park, S., 2021. Techno-economic analysis of producing xylo-oligosaccharides and cellulose microfibers from lignocellulosic biomass. *Bioresource Technology*, *340*, p.125726.

640312 Multiphase operations in the Taylor-Couette Disc Contactor

Abstract:

Multiphase reactions on industrial scale provide access to efficient and economical production of bulk chemicals. Nevertheless, the selection of reactor types is small and the modeling of fluid mechanical phenomena with more than two phases is difficult. This study focusses on the investigation and application of multiphase operations in the Taylor-Couette Disc Contactor (TCDC). The hydraulic characteristics and scalability have been investigated and confirmed [1]. It has been shown that the TCDC can continuously manage up to 4 phases simultaneously. Phase combinations are investigated by determining the holdup of dispersed phases and residence time distribution related to the rotating speed of the shaft and volumetric flowrate of phases. The TCDC is best modelled with the continuously stirred tank reactor cascade model. The results of selected phase combinations and their application will be presented. Continuous liquid-liquid-solid contact was successfully validated with the example of heterogeneously catalyzed esterification of acetic acid in aqueous solution and simultaneous extraction of the ester into the solvent phase, raising the total equilibrium conversion from 15% (without extraction) to 30%. Esterification experiments were performed and based on this data a reactor design algorithm was developed. To demonstrate the applicability of the Taylor-Couette Disc Contactor in gas-liquid operations continuous countercurrent neutralization of 0.1 molar sodium hydroxide solution with carbon dioxide was performed. The neutralization reaction was investigated for different rate of rotation, temperature and gas flowrate. Reaction rate and process efficiency were modeled based on film theory combined with a continuously stirred tank cascade approach. Neutralization of sodium hydroxide was performed until pH 9 was obtained to ensure neutralization reaction over the full reactor height. The experimental results confirm stable operation of the Taylor-Couette Disc Contactor in gas-liquid reactions. A simple column design algorithm for application in waste water neutralization with carbon dioxide is available. The hydraulics of four phase flow was investigated and confirmed. Applicability has not yet been confirmed, but will be some kind of heterogeneously catalyzed gas-liquid reaction with liquid-liquid phase transfer.

[1] A. Grafschafter, (2019), Simple and Effective: The Taylor-Couette Disc Contactor for liquid-liquid extraction (Design and Performance), PhD Thesis, TU Graz

639640 Energy Efficient Design of Industrial Processes through Life Cycle Cost Optimization

Abstract:

Industrial sectors account for about one third of global energy consumption which rely on fossil fuels as the primary source of energy. A large portion of the energy consumption is spent on production of utilities for the process plants to produce energy for other sectors. Due to the impacts of fossil fuel combustion on the environment lots of effort are made to minimize it and also environmental regulations are affecting the cost of energy to produce it. Furthermore, this is competitive world and with ever increasing energy prices the industrial communities are invited to conserve energy resources. It is common to find options for process changes in a production process and identify opportunities to reduce the operating and capital costs. Designing any system have many constraints and the challenge is to utilize the imposed constraints in a way to improve the system efficiency. Proper utilization of system design constraint along with the available technology can lead to efficient system design which provide many options for optimization. All system design is defining its efficiency by operation, the best operation can achieve/approach to the design efficiency but you cannot improve further without changing the system design. So, if a system is designed for higher efficiency will perform always superior that an inefficient designed system. Hence, it is very important to optimize the new facility design and develop methodology to investigate all possible alternatives during the design stage of the project. In this paper, optimization of the new facility design with simultaneous process and utility integration is demonstrated by conducting energy optimization assessment for one of the upcoming facilities during DBSP phase of the project. The objective of the assessment is to optimize the project life-cycle cost by assessing and examining all technical feasible design alternatives. The new facility is installation of additional units in the existing natural gas processing plant. Natural gas processing plants require a series of process units to remove contaminants such as water, carbon dioxide and hydrogen sulfide. These units have different process heating and cooling demands. Proper matching of the heating and cooling requirements leads to significant operating cost reductions. Excessive heat from process units can be recovered which may be used for satisfying heating requirements of other units or for power generation. The purpose of the new facility is to install dew point control unit, de-ethanizer and condensate stripper to meet the required products specification. The assessment results in energy and capital savings as well as GHG reduction and also assure that the new facility design is energy efficient while satisfying all intended objectives.

639228 Coupling Process Simulation and Computational Fluid Dynamics to Efficiently Design and Improve Multi-Scale Processes

Abstract:

Almost all real-world processes are multiscale processes, which mean that they have important features at different scales in time and/or space. Typically, not all scales can be resolved, so that for simulations of such processes certain assumptions or simplifications have to be made.

In the process industry for example, simulations are performed on a range of different scales and different purposes. Flow-sheet simulations are used for designing and validating the full process from raw material to the final product in terms of heat and mass balances, sizing and cost. Process simulations are conducted to get a deeper understanding of the underlaying chemical, physical or biological processes in unit operations, a basic step in a process. Such simulations are typically done with simplifying assumptions regarding the flow and other properties like ideal mixing in stirred tanks or plug-flow in tube-like devices and their focus is on the accurate description of the detailed physics. Finally, CFD simulations allow the detailed analysis of flow, turbulence and other relevant parameters in the equipment.

Since most processes depend, more or less, on local conditions, there is an obvious benefit to integrating process models into CFD. In practice, however, the applicability of this approach is limited mainly for the following reasons:

Depending on the system under investigation, the additional equations to be solved can generate a considerable overhead and lead to long computing times.

- The chemical-physical processes often happen on different time scales to the flow and mixing time scale.
- For design studies or variation calculations during design phase as well as during operation, results are required in a short time, usually in seconds or in a few minutes. This typically cannot be achieved with a CFD simulation.

Nevertheless, combining different scales can be very beneficial in terms of accuracy as well as getting a deeper understanding of the underlaying physics as previously shown e.g. for a wet granulation process [1].

In this contribution we show a new automated coupling method between the CFD tool Simcenter STAR-CCM+ and the process modelling tool gPROMS by PSE, where Simcenter STAR-CCM+ solves for all flow properties and gPROMS evaluates the fine scales based on the solution provided. The coupling between both codes is done in an automatic way. This includes the domain decomposition, the calculation of and the transfer of all relevant variables.

The details of the coupling as well as the superiority of this approach will be demonstrated for a packed bed reactor.

Packed bed reactors are one of the working horses in the chemical and process industry. They are used amongst others for highly exothermic or endothermic catalytic reactions like (partial-)oxidation or (dry, steam) reformation of Hydrocarbons, which is typically done in multi-tubular reactors. These types or reactor are used since several decades but there is still room for improvement of the performance of these reactors and rigorous simulations can help here to achieve desired goals with lower upfront invest. Recent advances in modeling particle-resolved packed beds [1,2,3] allows a detailed inside in the flow, species and temperature distribution in the beds and therefore also into the conversion of the surface reactions. Based on these simulations, model parameters for simplified 1D-models can be estimated [4].

References:

1. Eppinger, et al.; DEM-CFD simulations of fixed bed reactors with small tube to particle diameter ratios, https://doi.org/10.1016/j.cej.2010.10.053
2. G. D. Wehinger et al.: Detailed numerical simulations of catalytic fixed-bed reactors: Heterogeneous dry reforming of methane, https://doi.org/10.1016/j.ces.2014.09.007
3. N. Jurtz et al.: Advances in fixed-bed reactor modeling using particle-resolved computational fluid dynamics (CFD), https://doi.org/10.1515/revce-2017-0059
4. N. Jurtz et al.: Determination of Effective Transport Parameters from Particle-Resolved CFD Simulations for a Simplified Fixed-Bed Reactor Modeling (65a), AIChE Spring Meeting 2019, New Orleans

640328 CO_2 emission reduction by hydrogen based processing of mineral iron carbonate

Abstract:

On a global scale the iron and steel industry significantly contributes to carbon dioxide emissions. Direct reduction of mineral iron carbonate with hydrogen provides an alternative route to carbon based technologies for future sustainable, pig iron production. Direct reduction offers high potential for carbon dioxide emission reduction. For industrial implementation, sufficient supply of (renewable) hydrogen is crucial.

According to the state of the art technology mineral iron ore has to be upgraded by sintering before it can be fed to the blast furncace. In the sintering step iron carbonate (e.g. 3t of ore with an iron content of 33 wt%) is converted to hematite (e.g. 0.45 t Fe_2O_3) in oxidizing atmosphere plus 1.18 t of carbon dioxide related to one metric ton of pig iron. With CO_2 emissions from the blast furnace process the specific CO_2 load is about 2.2 t of CO_2 per metric ton of pig iron (or 1.9 t of CO_2/t pig iron when related to the $FeCO_3$ content of the ore). When alternatively mineral iron ore is processed in reducing atmosphere to form elemental iron instead of hematite ($FeCO_3 + H_2 = Fe + CO_2 + H_2O$) carbon dioxide emissions could be cut by 50% (except the energy demand for the smelter).

Therefore, direct reduction of $FeCO_3$ with hydrogen was investigated. With pure hydrogen from electrolysis complete conversion of $FeCO_3$ is obtained, arising the question, wether lean reducing gas (e.g. coker gas) might do the job as well.

The effect of the hydrogen quality and the impact of matrix constituents in the hydrogen source, such as methane and carbon dioxide, on the direct reduction of mineral iron carbonate were therefore investigated. It was shown that with hydrogen concentrations as low as 55 vol% complete iron carbonate conversion is possible. Carbon dioxide in the feed gas (27–63 vol%) favors magnetite formation. Methane in the feed gas (15–80 vol%) has no negative impact on the composition of the reductively calcined product. These results suggest that even low grade hydrogen sources are viable options for direct iron carbonate reduction. However, carbon dioxide emission reduction is limited to 50 % compared to the state of the art process. Advanced carbon dioxide emission reduction does need process gas treatment by using it for energy storage (actually hydrogen storage).

SPRING22
+18TH GCPS
A Joint AIChE and CCPS Meeting

Life Cycle Assessments: Corporate Value or Potential Risk Lies in the Details

Michael C. Stern
Exponent, Inc.
1075 Worcester Street, Natick, MA 01760
mstern@exponent.com

Jason James,[a] Ann Michelle Morrison,[b] Achim Wechsung,[c] Christopher Buehler[d]
Exponent, Inc.
a - 15375 SE 30th Place, Suite 250, Bellevue, Washington 98007
b – 1 Mill and Main Place, Suite 150, Maynard, Massachusetts 01754
c – 1075 Worcester Street, Natick Massachusetts 01760
d – 420 Lexington Avenue, Suite 1740, New York, New York 10170

Prepared for Presentation at
American Institute of Chemical Engineers
2022 Spring Meeting and 18th Global Congress on Process Safety
San Antonio, TX
April 10 – April 14, 2022

Z210111164847 - 9205

Life Cycle Assessments: Corporate Value or Potential Risk Lies in the Details

Michael C. Stern
Exponent, Inc.
1075 Worcester Street, Natick, MA 01760
mstern@exponent.com

Jason James, Ann Michelle Morrison, Achim Wechsung, Christopher Buehler
Exponent, Inc.

Keywords: Carbon Management, Product Design, and Sustainability Metrics for Energy Systems

Abstract

Life cycle assessments are an important tool to help companies measure, document, and communicate the potential benefits of processes and products. However, life cycle assessments are complicated analyses that rely on a variety of assumptions, databases, and inputs that may impact the applicability and reliability of the results. While firms that offer these services will often perform them in accordance with recognized standards such as ISO 14044, companies may not always be aware of the limitations, uncertainties, or assumptions underlying an LCA, even when performed to an ISO standard.

1 What is a Life Cycle Assessment (LCA)?

As companies plan to reduce the environmental impact of their operations, they are increasingly relying on tools such as life cycle assessments (LCAs) to understand the environmental impact of their products, processes, facilities, and technologies. A number of firms offer these services and generally will perform them according to recognized standards such as ISO 14044. Firms that specialize in LCAs often have access to large Life Cycle Inventory (LCI) databases (e.g., Ecoinvent) that allow them to assess the impacts of many processes based on the inputs and outputs of the system if LCI data is available for the inputs and outputs.

LCAs evaluate many parameters, including greenhouse gas (GHG) emissions, environmental impacts (e.g., ocean acidification), ecological impacts, human health impacts, and resource depletion (e.g., water, metals, or fossil fuels). Companies, however, may not be aware of the limitations, uncertainties, or assumptions underlying an LCA, even when performed to an ISO standard. For example, an LCA completed to the ISO standard may not be sufficient for use in marketing materials or regulatory disclosures unless further evaluated by a review board to meet the requirements of an Environmental Product Declaration (EPD).

2 Why do an LCA?

Z210111164847 - 9205

LCAs can provide many benefits to companies who want to internally assess or externally communicate the environmental impacts or benefits of their product. If a company either voluntary chooses to limit their carbon emissions or is regulated by a cap and trade or other regulatory mechanism, LCAs may assist the company evaluate whether to replace feedstocks, energy sources, or process technology to optimize their facility and processes to best meet their business objectives.

LCAs also allow companies to communicate their products' potential risks and benefits with stakeholders (customers, clients, investors, and insurers), and having an LCA performed by a third-party may add further credibility to the assessment in the eyes of stakeholders. However, as described below, LCAs can be subject to a variety of pitfalls that could provide misleading results or confusion if the limitations of the LCA are not properly understood and communicated.

3 What are the risks of an LCA?

LCAs require scientists to define the spatial temporal, and in some cases, abstract accounting boundaries of the life cycle and make numerous assumptions on a range of parameters used in the calculations. As a result, it is important to examine how an LCA is parameterized to understand its limitations and applicability.

A classic example of conflicting LCA results is whether corn ethanol generates less net GHGs than burning gasoline.[1] According to the U.S. Department of Agriculture, if one assumes that an ethanol plant uses coal for the heat required to convert corn to ethanol, the net GHG emissions for corn ethanol are greater than gasoline. Conversely, if that heat is derived from natural gas, corn ethanol provides a reduction in GHG emissions, demonstrating how different assumptions can reasonably lead to substantially different conclusions about the sustainability of a new technology.

Choice of LCA Boundaries

The boundaries of an LCA will substantially impact the result. Typically, the choice in boundaries focuses on which indirect (also known as Scope 3) emissions to include. Indirect emissions occur both upstream and downstream of a processing plant and can involve the emissions related to feedstock production, transportation, product use, and recycling or disposal.[2] LCAs are often performed to "cradle-to-gate," "cradle-to-grave," or more recently, "cradle-to-cradle." A summary of how these terms are typically used is shown in Table 1.

Table 1. Summary of typical terms used to describe LCA boundaries

Term	Meaning
Cradle-to-gate	From resource extraction until it leaves the factory gate
Cradle-to-grave	From resource extraction until, production, use, through disposal
Cradle-to-cradle	From resource extraction (recyclable or not), production, use, through recycle into a feedstock for another, or the original, product

Unfortunately, understanding whether an LCA is cradle-to-grave or cradle-to-gate may still be insufficient to be able to fully understand the scope and boundaries. For example, the LCA may or may not consider the contributions from constructing fixed infrastructure (also referred to as capital goods). In other words, some LCAs may only consider the emissions related to feedstocks and energy sources while others will also include the emissions of the warehouses, equipment, and other long-term assets required for operation. To fully assess and compare the results of LCAs, the precise boundaries and scope need to be clearly defined and understood because erroneous conclusions can result from comparing the results of LCAs without ensuring that the boundaries are comparable.

LCA Assumptions

When assessing the emissions resulting from a given feedstock, an LCA may rely on databases that provide typical emissions factors for the production of chemicals, carbon intensity of electricity, and other parameters. But these values may not be accurate for a given supplier or even the specific chemical being considered. The production of chemicals can often occur by multiple processes with varying emissions treatment systems that lead to large variations in emissions factors. Some chemicals may not have *any* specific data available to include in an LCA and may be approximated using data for other chemicals, which may misrepresent the emissions associated with the final product.

Choice of Energy Metric

Energy metrics, such as cumulative energy demand (CED), cumulative energy use (CEU), and others, may or may not be included when reporting the energy intensity of a product or process.[3] CED and CEU are both measurements of energy intensity, but CED includes the energy of materials, heat, and electricity sources, whereas CEU only considers the heat and electricity but does not consider the energy stored within the materials.

As a result of the various metrics used, the energy intensity reported by an LCA may not actually correlate to the required energy input of a product or process because other factors, such as the combustion energy of the fabrication materials (that are not combusted), are being included in the overall energy footprint. When using an LCA for decision-making, marketing, or disclosures, companies may want to consider the meaning of the energy metrics used in a supporting LCA to determine whether they are appropriate.

Comparisons to Current Industry Emissions

A comparative LCA, by definition, compares two processes or products. Oftentimes, comparative LCAs are used by companies developing new or novel processes and products to highlight their environmental benefits over existing technologies, which often rely on fossil fuels and petrochemical feedstocks. While these companies typically have reliable data for their own systems, the data for the conventional technology may not be as detailed and could lead to an unreliable comparison. The carbon footprint for any product or process (conventional or not) depends on how, where, and when it was produced. Consequently, comparative LCAs relying on outdated conventional process data or data generated in different regions may misrepresent carbon intensity, which could lead to allegations of "greenwashing." For example, significant disputes have arisen recently around the carbon footprint of blue hydrogen, highlighting the sensitivity of LCAs to the assumptions of the investigators.[4]

Companies interested in improving the sustainability of their operations have numerous choices for reducing their environmental impact, and all of these options may not be well represented in a comparative LCA produced by a technology developer. For example, instead of replacing a petroleum feedstock with a bio-based feedstock, the company could replace a hydrogen feedstock derived from natural gas with green hydrogen generated from renewable energy, electrify previously fossil fuel powered equipment, or install a system to capture and sequester carbon. Companies should not feel limited by the specific scenarios provided in a comparative LCA but rather may want to consider other options that provide similar environmental benefits and, potentially, additional business-relevant benefits. LCA studies also often consider water use, acidification potential, smog potential, ecotoxicity, and many other impacts that may present important tradeoffs or co-benefits that may be considered in addition to a reduced carbon footprint.

4 Conclusions

An LCA generated per ISO standards can provide a robust assessment of the environmental and health impacts of a product or process. However, the reliability of the assessment is based largely on the quality of the underlying assumptions and the choice of boundaries. The development and validation of many of these assumptions is the responsibility of the firm commissioning the LCA, who must provide certain system-specific technical knowledge to the firm they hire to perform the LCA.

When comparing to other systems, products, or technologies, the relevance of the comparison can be compromised if robust information is not available for competitor systems. Companies reviewing technology proposals and claims based on comparative LCAs may need to critically assess not only the claims made of the system being advertised but also the accuracy of the comparison to existing technologies.

5 References

[1] Gallagher PW, Yee WC, and Baumes HS. 2015 Energy Balance for the Corn-Ethanol Industry. US Department of Agriculture. February 2016. https://www.usda.gov/sites/default/files/documents/2015EnergyBalanceCornEthanol.pdf. Accessed April 2, 2022.

[2] GHG Protocol. Corporate Value Chain (Scope 3) Standard. https://ghgprotocol.org/standards/scope-3-standard. Accessed April 2, 2022.

[3] Arvidsson R, Svanström M. A Framework for Energy Use Indicators and Their Reporting in Life Cycle Assessment. *Integrated Environmental Assessment and Management* 2015, **12(3)**: 429–436.

[4] Howarth RW, Jacobson MZ. How green is blue hydrogen?. *Energy Sci Eng* 2021, **9**: 1676–1687.

639057 Applying Refining Fractionation Knowledge in Biomass Processing

Abstract:

It can be beneficial to separate complex mixtures into different fractions prior to further thermochemical processing. For example, significant advantages are obtained by separate processing of each fraction, obtained from the distillation of crude petroleum, through appropriate process such as Distillate Hydroprocessing, Fluidized Catalytic Cracking, Residue Hydrocracking, Visbreaking and Coking. Unlike the situation in refining, fractionation has still not found extensive application in biomass processing commercial technologies. In this presentation, challenges and potentials of using various fractionation techniques such as distillation, extraction, reactive extraction and leaching in processing biomass feedstocks and biomass derived intermediates will be addressed leveraging related knowledge from the refining industry. In addition, directional guidelines for future research for applying these technologies in biomass processing will also be discussed.

639660 Safety Challenges and Objectives in Rapid Lab to Plant Scale-up

Abstract:

Over the course of two years, Solugen Inc. went from lab proof-of-concept studies to building and starting up the world's first carbon-negative chemicals plant, using a proprietary chemienzymatic process. This talk will detail the inherent challenges in keeping a compressed timeline scale up safe and hazard free, from job walks on new laboratory techniques to PHAs and PSSRs on a first of its kind plant.

639229 Revalidating Risk Assessments in Laboratory and Pilot Plant Facilities: A Model

Abstract:

By its very nature, a research and development facility is in constant change. We are always trying something new, whether that be processes, chemical formulations, or production techniques. We manage our changes and investigate our incidents, but how do we ensure that we remain aware that all of the risks are managed and reduced to as low as reasonably practicable? See how the Shell Technology Center Houston is revalidating our risk assessments to ensure that we maintain our requirements to manage health, safety, and environmental risks.

639692 Pilot Testing for Liquid-Liquid Extraction Feasibility

Abstract:

Often times during the preliminary stages of process development for a potential liquid-liquid extraction application, a company is looking to determine if the process is feasible without using a lot of resources, time or materials. In previous papers, a KARR® Column Bench Top Unit (BTU) was identified as a laboratory column to be used for this purpose (1). However, the BTU is limited by an air drive motor to operate at agitation speeds above 100 SPM. Many of the new applications, specifically ones involving extraction from biobased feed streams require lower agitation speeds in order to operate effectively without emulsifying or flooding the column. A test method using a 5/8" diameter x 4' plate stack height KARR column has been developed that can be used for these more difficult extraction processes. This paper will present case studies where this size column was used to demonstrate the feasibility for three (3) different applications.

639666 Grams to Tons: Rapid Scale up of a First of Its Kind Chemienzymatic Process

Abstract:

In the Fall of 2021, Solugen started up the world's first carbon negative chemicals plant, using a novel chemienzymatic process. This talk will outline the journey Solugen took from benchtop to pilot to manufacturing scale, with emphasis on process scale up, as well as on the design, installation, and start up of a 10 kiloton per annum plant in a two-year timeline from start to finish.

639981 Efficient and Sustainable Manufacturing of MOF-Polymer Composites

Abstract:

Over the past decades, significant research has shown that manufacturing MOF-polymer composites is a promising strategy for improving its practical adaptability in storage, separation, and catalysis applications since unshaped MOF powders are not appropriate for deployment in engineering processes. However, there are some gaps between laboratory-synthesized and commercial-deployed MOF-polymer composites: most MOFs are synthesized in lab in milligram/gram scale under batch reactions and are then used to manufacture composites in a discontinuous process; it requires a large volume of hazardous solvents during the solvothermal synthesis, which posts more environmental issues and increases material costs.

In this work, we propose a continuously one-step, rapid, and solvent-free approach to manufacture MOF-polymer composite directly from solid reagents (polymer matrix, metal irons, and ligands) via extrusion that can reach a few kilograms per hour scale compared with conventional method in grams/batch/day. As a result, this approach allows MOF-polymer manufacture process to circumvent the traditional multi-step, low scalability efficiency, high solvent demand, and time and energy consuming methods. For the first time, this environmental-friendly approach has been demonstrated in a case study of ZIF8-polypropylene composites using Process 11 parallel twin-screw extruder. A further study on different polymers with various MOFs are also performed to validate this idea. The effects of the extrusion parameters on the MOF dispersion efficiency and scale-up risk analysis within the polymer matrix will be studied in detail to investigate an optimal manufacturing condition.

639291 Success Stories and Lessons Learned on Safety and Operational Excellence at Pilot Plants

Abstract:

Pilot plants are a critical de-risking step for scale-up of process technologies. But, have you ever wondered how pilot plants command special considerations for safety and reliable operations due to the intermediate scale involved?

This presentation will cover a broad range of success stories/lessons learned for managing safety and achieving operational excellence at pilot plants from multiple projects for process development at Shell Oil Company over the last few years. Specifically, some of the best practices that were beneficial in navigating the unprecedented challenges posed by pilot plant operations during the COVID-19 pandemic will also be discussed.

639185 Pilot Plant Design for Biological Systems in Heavy Industry

Abstract:

Cemvita Factory is a leader in synthetic biology for application in heavy industries. As part of our ongoing work to reverse climate change we have built out pilot plants for biological systems up to 1000 liters. This presentation will describe the challenges, lessons, and insights that building and operating these plants has provided. The presentation will address plant design, operation, data collection, and process optimization.

639048 Development of Intensified Continuous Processes: Lessons Learned through Scale up of New H_2s Scavengers

Abstract:

The presence of naturally occurring hydrogen sulfide (H_2S) is found in oil and gas reservoirs worldwide. There are several options for the removal of H2S from the production of oil and gas and these options range from fixed assets that scrub H2S to chemical injection at the wellsite. The majority of the non-reversible chemical H2S scavenger market is based on triazine technology derived from the reaction products of formaldehyde and amines (most common monoethyleneamine MEA). In recent past, there has been an active search by all to improve the overall performance of H2S scavengers in oil and gas production.

Major topics that the current technology can be improved upon include: increased consumption of H2S, reduction of nitrogen contamination of crude oil, reduction of scaling issue, elimination deposition of by-products, addressing EHS concerns and minimization of the disposal of products/by-products. Some of the biggest hurdles with new H2S scavengers are the following: ensuring fast kinetic reaction rates, system compatibility, similar molar ratio (i.e., consumption rates), minimal precipitation of scavenger/by-products, large manufacturing scale, large raw material supply and price points that allows for a benefit of change. Ideally, the new product would offer better consumption performance versus the incumbent, reduction in nitrogen content and minimization of deposition.

A product identified over 20 years ago as a potential replacement was 2-Methyl-3-buten-2-ol (MBO). However, MBO has had limited application in field until recently. MBO is a reaction of amine (monoisopropylamine) and formaldehyde which offers some of the same benefits of triazine but improves upon the areas of increasing consumption of H2S per mole of scavenger, reducing nitrogen content in crude oil, reducing the by-product deposition potential and is already produced on large manufacturing scales. Production on a large manufacturing scale is critical for success of any H2S scavenger into the oil and gas market.

The body of work performed here demonstrates Lubrizol's approach to rapidly scaling H2S scavenger manufacturing production through modular process intensification and conversion of a batch process to a continuous reaction process. The process described herein involves the reaction of formaldehyde with an amine resulting in the formation of water as a by-product of reaction which subsequently needs to be removed to drive conversion to the desired product. Despite the apparent fast reaction rate between formaldehyde and amines, the current batch process has a standard processing time of 18 hours. This long time cycle is due to several factors including: bulk solid paraformaldehyde charging, slow heat up and cool down times, and mass transfer and equilibrium limitations during the reaction and stripping stages of the process. In an effort to debottleneck manufacturing and improve product quality, the project team sought to convert the batch process to a modular continuous process. Initially, a reaction rate and calorimetry study were conducted to elucidate the reaction kinetics and overall heat of reaction. This work became the foundation for continuous reactor train design to maximize heat removal and control the reaction temperature to minimize by-product formation. A subsequent study involved the development of a novel reactive distillation design which enabled a single pass, continuous stripping process. The resultant product is an analytically equivalent product to the batch process.

The project team is currently focusing on scaling the new continuous process in a modular shipping-container-size design for rapid deployment in production. The total process development timeline for this work was less than 10 months and the total time from inception to the deployment of a commercial unit is expected to be less than 24 months.

639196 Corn Bale Deconstruction Using Physical Techniques for Improved Anatomical Separation

Abstract:

Corn stover, a mixture of various anatomical fraction of corn plant, often comes with introduced inorganics from the harvest and collection process. The variability of corn stover can be reduced with a combination of size reduction (knife mill, hammer mill etc.) and size separation (air classification, sieving etc.). The combined process, incorporating different air speed and sieving, can generate a fraction of stem & cob each with 75% purity and yield. The physical techniques can also reduce ash content by removing (30%) introduced inorganics. Moreover, low ash content (<7%) husk, sheath and leaves were also generated. The separation of these fractions enables fraction specific processes including reduced cost of pretreatment and fiber-based coproducts.

639260 Key Methods and Strategies to Accelerate Speed to Market and Reduce Cost and Risk for Scaleup of Chemical Processes

Abstract:

Transitioning lab-scale chemical processes to medium and full-scale commercial production can be costly, time-consuming, risk-intensive, and cause inefficient use of key resources.

Following a systematic method of defining requirements and specifications for commercial production, along with making proper decisions at key project milestones, will help you assure that minimal expenditures will be made for processes that are not good for scaleup and good opportunities will be brought on-line efficiently. Areas of greater unknowns and/or risks will be identified and targeted.

In this presentation, we will describe a method to follow when engineering and designing for scaleup of chemical processes, along with evaluating the economic viability of the scaleup at key project milestones. We will also share examples of actual projects that have been completed and others that are under way. In addition, we will describe a way to monitor initial and ongoing processes and products in a way to minimize startup time, product variability and costs.

We will discuss means and strategies for collaborative scale-up processes and execution with third-party partners with relevant process experience. The interchange, and potential advantages and cautions of such relationships, will also be explored. Examples of getting to operationally-successful finished systems in desired timeframes following proven scale-up development processes will be presented.

Re:Build Optimation is a full-service engineering, design and skilled trades firm that works with clients to complete projects involving phases from process development through design, installation and commissioning.

640533 Life Cycle Analysis on the RENUVA™ Mattress Recycling Program in Europe

Abstract:

Every year, ~30 million mattresses are thrown away in Europe. By using chemical recycling of PU foams from mattresses to generate RENUVATM polyols, we can give a second life to this valuable material without performance tradeoffs. Dow's RENUVATM mattress recycling program is an innovative solution to increase circularity for end-of-life PU products in collaboration with the value chain. As part of this innovation partnership, Eco-mobilier will employ its used mattress collection and dismantling capabilities to supply post-consumer polyurethane foam to Orrion Chemicals Orgaform's (OCO) chemical recycling unit in Semoy, France. As the program is moving towards commercialization, an in-depth Life Cycle Assessment (LCA) analysis was performed to better understand the benefits for society and enable commercial claims following an external review by the sustainability consulting firm Quantis. LCA is a science-based method of choice for examining the total environmental impact of a product or service. The goal of this study was to determine cradle-to-gate environmental burdens of conventional Dow-produced VORANOL™ 3322 polyols and understand how it compares on a functional unit basis with a new Dow product, namely RENUVATM polyols, that incorporates polyols from recycled polyurethane foam (PU) mattresses. The study considered the potential environmental impact categories included in the EF Method 3.0 (adapted) V1.00, widely accepted in the European Union (EU). The benefits of this chemical recycling process are two-fold, i.e. both creates 2.4 kg of polyols for use in new mattresses' PU foam AND treats 1 kg of end-of-life (EoL) PU foam from old mattresses. It is the combination of these two benefits that defines the functional unit used for this study. The cradle-to-gate LCA results demonstrate that RENUVATM polyols offers the potential for a significant reduction of the environmental footprint associated with polyols. A key observation was the 54% reduction of climate change impact relative to incumbent technology for the given two-part functional unit (4.77 kg CO2 eq for RENUVATM vs. 10.3 kg CO2 eq for incumbent). One trade-off in environmental impact categories was identified relative to ionizing radiation. However, it is of moderate impact once weighted against the main ones affecting the EU region and is specific to the high share of nuclear power in the French electricity grid.

640158 Technology Advancements for Producing Biodiesel from Waste Lipids

Abstract:

Starting with the passage of the Renewable Fuel Standard (RFS) in 2005 and more recently a societal and customer demand for low carbon fuels, waste lipids have been an increasing source of raw material used for producing biomass based diesel. These feedstocks, such as used cooking oil and distillers corn oil from the ethanol industry, have presented new opportunities and processing challenges for biodiesel manufacturing. A case study will be presented to show how Renewable Energy Group developed the technology and capabilities to become a leader in utilizing these waste lipids, thereby supporting a circular economy.

Renewable Energy Group, Inc. ("REG") is leading the energy and transportation industries' transition to sustainability by converting renewable resources into high-quality, sustainable fuels. REG is an international producer of sustainable fuels that significantly lower greenhouse gas emissions to immediately reduce carbon impact. REG utilizes a global integrated procurement, distribution, and logistics network to operate 11 biorefineries in the U.S. and Europe. In 2020, REG produced 519 million gallons, or 1.7 million metric tons, of cleaner fuel delivering 4.2 million metric tons of carbon reduction. REG is helping to meet the growing global demand for lower-carbon fuels and leading the way to a more sustainable future.

638724 Sustainable Energy Efficiency Improvements through Methodology of Prioritizing Selections

Abstract:

Industrial facilities are rapidly changing over the years, various technological advances have led to break-through its performance. The transformation is brought from new inventions, development of sophisticated processes/technologies, automation, digitalization, the list is continuously increasing. These technical development and strategies to realize it on the life-cycle basis forces the industries to explore sustainability concepts within the scope. On the other hand, the task to reduce Green House Gas emissions to protect environment is needed than ever before. Consequently, the revolution and evolution of products, processes, technologies, and resources, and sustainable requirement to mitigate climatic anomaly beams on the industrial sector and ordained it to cater an approach to optimize all efforts. The challenge is to develop a methodology by accommodating all the technological advances and sustainability requirements and transform facilities into a better performing asset with reduced emission. The solution is to make industrial facilities energy efficient which enable them to be competitive and environmentally friendly.

Saudi Aramco has never ending quest for innovation in technological future products and services. It is evident from its pledge to Operational Excellence and commitment to carbon circular economy. Saudi Aramco is resolute to develop and retrofit all of its facilities to be best-in-class from energy efficiency perspective which is palpable form its Energy Policy. To take on the challenge, Saudi Aramco developed new methodologies to design/retrofits industrial facilities to realize energy efficiency. The interplay of Energy Trilemma requirements i.e. techniques for energy savings, climate change abatement, and overall sustainable development is considered within the framework. Recent enthusiasm for using renewable energy resources, both to reduce expenses associated with traditional sources and to improve environmental conditions is overwhelming. But the emphasis to ensure optimal utilization of materials to achieve energy efficiency improvement through sustainable means is reflected in the developed approach.

In this paper, the developed technique is described which needs to be followed in retrofitting industrial facilities to make them the best-in-class. The methodology is conferred in the form of "Energy Pyramid" which is built on the principle of prioritization i.e. build the base first and then progressive layers are added until we finally get to the peak. It is likely that by following the approach, manufacturing industry of the future will become energy efficient and fully embrace the best practices to optimize resources utilization while consuming less energy. Industrial facility case studies are utilized to enrich and illustrate the subject and demonstrate the methodology appropriately to put them on track to achieve international climate and energy goals.

AUTHOR INDEX